I0486281

WARP SPEED A PLUS

NEWTONIAN RELATIVISTIC MECHANICS

Employing Cartesian Invariance with predicted variance of "the" speed of light in a Once Accelerated Reference System.:

By

Gary Mike Colasuono

© 2004 by Gary Mike Colasuono. All rights reserved.

No part of this book may be reproduced, stored in a retrieval system, or transmitted by any means, electronic, mechanical, photocopying, recording, or otherwise, without written permission from the author.

ISBN: 1-4033-1055-6 (e-book)
ISBN: 1-4033-9540-3 (Paperback)
ISBN: 1-4107-9889-5 (Dust Jacket)

This book is printed on acid free paper.

1stBooks – rev. 02/13/04

Table of Contents

HOW SPECIAL RELATIVITY VIOLATES THE PRINCIPLE OF THE CONSERVATION OF MASS AND ENERGY.:

A THOUGHT EXPERIMENT

BY

GARY MIKE COLASUONO

Gary Mike Colasuono

Abstract: By comparing two relative points of view we come up with an inconsistency and with an absolute point of view.:

Consider a relativistic space craft moving at .866c relative to the rest of the universe. From the point of view of the rest system of a stationary universe, the mass of the relativistic space craft approximately doubles from its rest mass. (Note.: $m = m_0(1-v^2/c^2)^{-.5000...}$). This can be accounted for by employing the Conservation of Mass and Energy and the Equivalence of mass and energy by realizing the additional mass originated because of the past thrust of the relativistic space craft which got the space craft up to .866c in the first place. That is, energy was converted into relativistic mass.

However, from the point of view of the reference system of the relativistic space craft, according to special relativity, the craft is at rest and it's the rest of the universe that is moving at .866c. According to the special theory of relativity the mass of the rest of the universe can be considered to have approximately doubled over its rest mass of $m_{0Universe}$.
According to the equivalence of mass and energy this represents a relativistic energy for the "moving"

universe of $2m_{oUniverse}c^2$, and therefore to an energy difference between a universe at rest and a universe moving relativistically at .866c of $m_{oUniverse}c^2$. This however presents a serious violation of the principle of the Conservation of Mass and Energy since there isn't any way to account for the origin of that vast amount of energy. Where did the initial energy input to the stationary universe come from to yield a mass increase of $m_{oUniverse}$, which represents the entire rest energy of the initial stationary universe? To answer this question satisfactorily, we must analyze the past history of both the space craft and the rest of the universe from both the point of view of a moving space craft and stationary universe and the other point of view of a stationary space craft and moving universe.

Now from both points of view there was an initial energy input to the relativistic space craft to account for the moving space craft's mass increase, that is.: $m = E/c^2$.: The space craft's initial thrust which got the space craft up to .866c in the first place.

Also and however from both points of view there was no corresponding initial energy input to a stationary universe to account for a mass increase of a now moving universe.

Conclusions.:

1.: The relativistic space craft, the object which contains the once added energy, is in a state of absolute motion relative to the stationary universe.

2.: The rest of the Universe, having received no energy input as did the relativistic space craft, is at absolute rest relative to the relativistic space craft and to itself.

3.: Special Relativity violates the principle of the conservation of mass and energy.

4.: The mass increase of the relativistic space craft is real and absolute with the space craft's initial thrust providing the energy input to account for its absolute mass increase.

5.: The energy content of the Universe both before and after the time the relativistic space craft was accelerated is $m_{oUniverse}c^2$, the rest energy of the stationary universe, since there was from both points of view no energy input to the stationary universe while the relativistic space craft was accelerating.: that is.: There was no initial energy input to the stationary universe to account for its mass to increase.

6.: The history of the events localizes the energy of the moving system.

7.: Moving and rest could be compared to a democracy with the major portion of the mass always being at or near rest.

8.: Newtonian Relativistic Mechanics most closely represents the truth concerning how nature really works.

9.: If you try to appeal to General Relativity you run into the problem of a violation of the Principle of Cause and Effect. That is.: What force could possibly account for the acceleration of an entire universe in one direction, and beyond that what energy source could account for the vast amount of energy needed to save Special Relativity?

10.: Final conclusion.: The relativistic space craft is in absolute motion relative to the rest of the universe, but the rest of the universe cannot in any circumstance be considered moving relative to the relativistic space craft. Motion is absolute in the Newtonian sense.

11.: The energy of a system is localized by the past history of the system in question.

Gary Mike Colasuono

GARY MIKE COLASUONO

ON

THE MATHEMATICAL PROOF OF WHY SPECIAL RELATIVITY IS LOGICALLY INCONSISTENT.:

Gary Mike Colasuono

FORWARD:

In the following analysis, I have taken the liberty to employ the Bohr and Bohr Sommerfeld theories of atomic theory. You might say, for the sake of argument, that these theories are outdated. However, consider, that the Bohr Sommerfeld theory correctly predicts precisely the correct atomic spectra lines in every case. That is, for all energy levels, and does this by matching up the correct atomic spectra lines with what has to be the correct energy level transitions. There is no mistake here, there is no coincidence here, and the predictions are precise and correct.

Even though Heisenberg's mathematics conceal the complete physical picture (which isn't given here) of what is happening, Heisenberg's mathematics in no way negates Bohr Sommerfeld physics. In other words, Heisenberg's mathematics in no way invalidates Bohr Sommerfeld physics. Heisenberg's mathematics merely takes additional factors into account and at the same time and in the process conceals the physical picture that Bohr Sommerfeld correctly gave. Otherwise, how do you explain away the precise and correct predictions that

Bohr Sommerfeld gave in their theory and so subsequently won the Nobel Prize for?

Even if you say the Bohr Sommerfeld argument is completely devoid of logic (in that case, whatever you consider the word logic to signify), the following analysis can be performed on an ideal hypothetical sledge hammer (with all its mass in the hammer head) rotating about the end of its handle in a once accelerated reference system, a spaceship if you will, which originated from a non once accelerated reference system, earth if you will. According to Newton's law of inertia, angular momentum will also be conserved for this gross rotating system. The author much prefers the elegant analysis which follows and does employ the Bohr and the Bohr Sommerfeld pictures of physical reality. However, if you insist upon being hard headed, you may employ an ideal hypothetical sledge hammer, and you will still reach my conclusion.

Finally, if you want my development of the quantum mechanics then you must first publish the papers I have already sent you.

In addition, if you say my logic is faulty (in any place), then it is your responsibility to physics to tell me where and how it is faulty (which is isn't), or else find yourselves another occupation.

Final note, the Bohr circular orbital case is sufficient to prove special relativity wrong, because there does exist a contradiction.

The analysis follows.:

Gary Mike Colasuono

GARY MIKE COLASUONO

ON

THE MATHEMATICAL PROOF OF WHY SPECIAL RELATIVITY IS LOGICALLY INCONSISTENT.:

Gary Mike Colasuono

16

1.: This mathematical proof is based on the physical principle of the Conservation of Angular Momentum for the specific and sufficient case of circular motion and can be logically generalised to the infinite number of cases of elliptical motion if you consider a conceptual generalisation of the conservation of angular momentum to the concept of the conservation of comparative and instantaneous angular momentum. If a particular point in an elliptical orbit is compared at two different points in time, where the mass is also different compared at the two different points in time (but remember: at the same point in space of a particular elliptical orbit), then the instantaneous orbital velocity for that particular point in space will vary precisely inversely as the mass if and only if the radius vector for that particular point in space remains a constant, and in general every instantaneous orbital velocity for every particular point in space contained in the elliptical orbit will vary precisely inversely as the mass (of a moving particle) if and only if the radius vector for that particular point in space contained on the elliptical orbit remains constant. The above is true if and only if the central force field is the only influence influencing the elliptical motion. That is.: the field affecting elliptical motion must be conservative. This general form of the Conservation of Angular Momentum is useful for studying and generalising the Bohr model of orbitals to the much more general Bohr Sommerfeld model of elliptical atomic orbitals. The above can be applied to the specific case of Bohr type orbitals.

2.: For circular Bohr type orbitals.:

The Conservation of Angular Momentum can be stated as follows.:

$$mV \times R = \text{a constant}$$

Note.: Keep in mind, that for a circular orbit the orbital velocity is related to the frequency of rotation by the following formula:

$$V_{orbital} = Cf$$

where C is the circumference and f is the frequency of rotation about the circumference.

3.: In special relativity the following relationships are held true:

$$m = m_o(1 - v^2/c^2)^{-.5000...}$$

$$f = f_o(1 - v^2/c^2)^{.5000...}$$

(The system being considered: an electron moving around nucleus in a spacecraft).:

Notice two things.:

A.: The above relationships are precise and have been experimentally verified.

B.: The "v" in the above relationships is a linear velocity and isn't, the author repeats, isn't the orbital velocity "V".

4.: If in 2 above, the physical law of the Conservation of Angular Momentum is true, and if m, the mass (of an electron say) increases to a particular and specific value, then $V_{orbital}$ has to decrease inversely, precisely inversely, to the increase in mass, if and only if the radius vector of the angular momentum remains constant. This has to be the case due to the experimentally verified relationship stated in 3 above of this paper.

(Note.: The circumference also has to be a constant due to the experimentally verified relationship stated in 3 above of this paper.).

5.: However, the Lorentz transforms demand length contraction when time dilation takes place. This directly contradicts 1, 2, 3, and 4 above, since we have already determined that the radius vector of the orbit in question must remain constant if special relativity's relationships in 3 above are true and precise as have already been experimentally verified.

6.: Given the specific case of a circular Bohr orbital, the author has shown there exists an internal logical inconsistency in special relativity which is a sufficient internal inconsistency.

7.: In the general case of elliptical Bohr Sommerfeld orbitals, we can look at a particular point on the elliptical orbit at two different points in time after a mass change, and again apply basically the same logic, and again arrive at the fact of an internal logical inconsistency in special relativity. (Note.: radius vector is a constant at same spatial point of ellipse.).

8.: This internal logical inconsistency is not explained by General Relativity theory, and cannot be explained by General Relativity theory.

Q.E.D.

9.: Aside.: The author believes Newtonian Relativistic Mechanics to be the correct representation of how nature works. There exists a deciding experiment.:

If the velocity of light varies in the following way, then Newtonian Relativistic Mechanics is most probably essentially correct.:

$$c' = c(1 - v^2/c^2)^{-.5000...}$$

where c' represents the velocity of light measured by the once accelerated moving reference system as compared to the velocity of light of the non once accelerated reference system, namely c.

In other words, the velocity of light depends on the ((causal) accelerated or non accelerated) history of the physical reference system.

Gary Mike Colasuono

NOTE.: Vector Products and ellipses.:

The mass, m, is a scalar. The magnitude of the instantaneous velocity vector varies precisely inversely as the mass, m, at every particular spatial point on the ellipse. The direction of the instantaneous velocity vector remains constant at every particular spatial point on the ellipse. Both the magnitude and direction of the radius vector remain constant at every particular spatial point of the ellipse. All of the above are true, if the shape is to remain an ellipse.

NEWTONIAN RELATIVISTIC MECHANICS

Employing Cartesian Invariance with Predicted Variance of "the" Speed of Light in a Once Accelerated Reference System.:

Author.: Gary Mike Colasuono

A Relativistic Abstract.: Newtonian Relativistic Mechanics is based on the premise that there are four separate effects which influence a Relativistic Reference system (and can also be superposed on one another).:

1.: The Doppler effect due to relative Motion.

2.: The Time-Delay effect due to the finite speed of light.

3.: The Time-Dilation effect due to the conservation of angular momentum.

4.: General Relativistic like effects.

This paper is primarily concerned with the Time-Dilation effect due to the conservation of angular momentum.

Motivation for a new Theory.:

A need for understanding interactions along the Time Axis and the need for the correct concept of varying degrees of motion along the Time axis.: BOTH of which neither Special Relativity nor General Relativity provides. The correct concept of varying degrees of motion along the Time Axis leads to a Unified Field Theory of Gravity based on electromagnetism properly (The mathematics based on conceptualization) Generalised to four dimensions.

FUNDAMENTAL POSTULATES.:

1.: The speed of light depends on the past history of the reference system and the rate of clock rotation in the reference system in question.

2.: The rest reference system depends on and is determined by the locally preferred Gravitational reference system (the predominant Gravitational fields, if you will.). THIS accounts for the null result of the Michelson-Morley experiment if you think of the *locally preferred* Gravitational fields as the "Aether" of old.

3.: THE *Once Accelerated Reference System is the moving reference system.*

Philosophical Basis of this Theory.:

There is "memory" (Real Mass Change) in Physics, in terms of energy due to acceleration having been converted into mass increase in the process of mass having been accelerated ($E = s \cdot am$ and $m = E/c^2$ which signifies $m_{increase} = s \cdot am/c^2$). This "memory" (Real Mass Change due to acceleration) shows up as Time-Dilation and a Variation in the speed of light. (Note.: s is displacement, a is acceleration, m is mass, E is the energy due to acceleration, and c^2 is the speed of light squared. (Decrease shows up if deceleration which signifies the process is reversed with mass being converted back into free energy.)).

Derived Conclusions.:

1.: Cartesian Invariance in three dimensions (Newtonian absolute space and spatial invariance).:

$$A^2 + B^2 + C^2 = A'^2 + B'^2 + C'^2 \qquad (1)$$

Generally.: $S^2 = A^2 + B^2 + C^2$.: AS an extension of the Pythagorean theorem of two dimensions valid for all coordinate systems.

2.: Cartesian Invariance in four Dimensions.:

$$A^2 + B^2 + C^2 + c^2T^2 = A'^2 + B'^2 + C'^2 + c'^2T'^2 \qquad (2)$$

Subtracting (1) from (2).:

$$c^2T^2 = c'^2T'^2$$

Taking square roots.: $cT = c'\,T'$

Getting.: $c' = cT/T' \qquad$ (2A)

Let.: $M_T = T/T'$ so $c' = cM_T$

3.: The Equivalence of mass and energy.:

Following Albert Einstein and knowing the need for a correct mass-energy formula which gives classical results for slow-moving and non-moving masses.:

$$m_o(1 - v^2/c^2)^{-.5000...} = m_o(1 + (1/2)(v^2/c^2) + (3/8)(v^4/c^4) + ...).$$

The above series converges rapidly when v is small, so we need only be concerned with the first two or three terms. If we multiple by the constant c^2, then the first term in the second member is simply the rest energy.: $E = m_o c^2$, and the second term in the second member is simply the kinetic energy. The third and further terms of the second member are further corrections on the kinetic energy. So fitting Relativistic Results with classical limits we see the correct mass increase Relationship.:

$$m = m_o(1 - v^2/c^2)^{-.5000...} \qquad (3)$$

4.: Let's go another step further than Albert Einstein and derive time dilation from mass increase.:

The Law of The Conservation of Angular Momentum States.:

mVr = a constant = 2 "pi"fmr^2

where f = frequency and 2 "pi" f = "omega"

IF r = a constant. Then as m obeys (3) and increases.: f, frequency, decreases in inverse proportion.:

$$f = f_o(1 - v^2/c^2)^{+.5000...} \qquad (4)$$

Then replace f with T' and f_o with T and start with any convenient origin.
Note.: See Appendix A
Note.: The above argument works equally well with quantized angular momentum for every given atomic system.

5.: Absolute Time Dilation.:

Let $A = (1 - v^2/c^2)^{+.5000...}$

Then.: T' = TA

And   T = T'/A

Is the definition of Absolute Time Dilation and compare with special relativity and its relative time dilation.:

T' = TA

And

T = T'A

6.: Putting equations (2A) and (4) together we finally get.:

$$c' = c_0(1 - v^2/c^2)^{-.5000...} \tag{5}$$

where c' is the speed of light a once accelerated (moving) space craft would observe.

This is one prediction of Newtonian relativistic Mechanics which distinguishes Newtonian Relativistic Mechanics from special relativity and is a prediction which hasn't yet been tested.

Note.: In a once accelerated non-accelerating reference system with no forces acting there on, length contraction would violate Causality, so for the sake of maintaining causality.: There is no length contraction connected with Newtonian Relativistic Mechanics. This is yet another distinguishing factor.

7.: Time Motion on a general basis on an atomic scale can be stated simply as $M_T = T/T'$. (See derived conclusions, Section 2) For Unified Field Theory Time Motion is more specific and is stated on a General basis on a Nuclear Scale.

8.: Absolute space.: $v' = vM_T$

Where v' is what the moving (Once Accelerated) reference system observes and v is what the rest reference system observes. (Note Sections 2, 5, and 7). Keep in mind with Newtonian Relativistic Mechanics.: Real Space Distances (vT and v'T') are Absolute.

NOTE.: RELATIONSHIPS.: $vT = v'T'$ and $v^2/c^2 = v'^2/c'^2$.

9.: The energy of a system is localized by the past history of the system in question. (Once Accelerated for example.).

RELATIVISTIC NUMBERS

BY

GARY MIKE COLASUONO

Gary Mike Colasuono

1.: $M_T = T/T' = (1 - v^2/c^2)^{-.5000...}$

2.: $v' = vM_T$, $v = v'/M_T$, $c' = cM_T$, $c = c'/M_T$

3.: $(v^2/c^2) = (v'^2/c'^2)$

4.: $c' = c(1 - v^2/c^2)^{-.5000...}$

5.: $c' = c(1 - v'^2/c'^2)^{-.5000...}$

$c'(1 - v'^2/c'^2)^{+.5000...} = c$

$c'^2 - v'^2 = c^2$

$v'^2 = c'^2 - c^2$

$v' = (c'^2 - c^2)^{+.5000...}$

6.: $v' = (c^2M_T^2 - c^2)^{+.5000...}$

$v' = c(M_T^2 - 1)^{+.5000...}$

7.: Then Let $n = (M_T^2 - 1)^{+.5000...}$

Then.: $M_T = (n^2 + 1)^{+.5000...}$

Gary Mike Colasuono

8.: v' = nc, where n = "RELATIVISTIC MACH NUMBER"

9.: $v = (n/M_T)c = (((M_T^2 - 1)^{+.5000\ldots})/M_T)c$ (9A)

$v = (n/(n^2 + 1)^{+.5000\ldots})c$ (9B)

10.: Multiply both sides of equations (9A) and (9B) by

$$M_T$$

Getting.: $v' = (nc) = (((M_T^2 - 1)^{+.5000\ldots})/M_T)c'$
(10A)

And.: $v' = (n/(n^2 + 1)^{+.5000\ldots})c'$ (10B)

Note.:

v = The velocity that the stationary non once accelerated observing system measures for the moving system (the once accelerated system). This is due to the lockstep fashion through time determined by the observing systems rate of motion through time.

c = A specific case of "v". (Same lockstep fashion through time determining the value of "c".: that is.: c = 299,792,458 meters *per earthship second*.)

v' = Velocity the moving system (The once accelerated system) measures for the moving system's own velocity and is the produce of "v" and the moving system's time motion.

c' = value of speed of light the moving system (once accelerated system) observes and is the product of "c" and the moving system's time motion. c' can take on all values between zero and infinity.

The Test.:

The Planet Mercury Project.: To land an Atomic Clock on the surface of the Planet Mercury.

$$c" = c_{Earth}((1 - v_E^2/c^2)^{+.5000...})(1 - v_M^2/c^2)^{-.5000...}))$$

c" = The speed of light on the surface of the planet Mercury as seen from the surface of the planet Mercury.

v_E = The orbital speed of the earth around the sun as seen from the earth.

35

v_M = The orbital speed of the planet Mercury around the sun as seen from the Earth.

c = The speed of light on the surface of Earth as seen from the Earth.

$$c' = c_{Earth}(1 - v_E^2/c^2)^{+.5000...}$$

c' = the speed of light on the sun's surface.

Final conclusion.: The faster we move through space, the faster we move through time, the faster the speed of light becomes.

References.:

1.: Bohm, David, *Special Relativity*

2.: Eddington, A.S., *Theory of Relativity*, Cambridge University Press.

3.: Feynman, R.P., *The Feynman Lectures on Physics*, Volumes I and II and III.

THE ORIGIN OF ROTATION AND SPIN IN THIS UNIVERSE

BY

GARY MIKE COLASUONO

Gary Mike Colasuono

Abstract. Everything has an origin, not the least of which are rotation and spin.

Background.:

AS you know, there are stars, then star systems, then galaxies, then groups of galaxies, then "the" Universe. What you may have already guessed is that there maybe also groups of Universes and also an infinite number of groups of Universes, which the author terms: The Hyperuniverse.

This Hyperuniverse would then be infinite in terms of both time and space.

Up until this paragraph you may think to yourself, this is all conjecture. However, consider the fact that there are cosmic rays which are so energetic, no known nuclear process, including time reversal, can possibly account for their energies.

However. There is another non-nuclear possibility.: The Doppler Shift.

Consider two or more rapidly expanding Universes, each expanding around their own centers and thereby rapidly expanding towards one another.

These universes are then very rapidly coming together with the information from one universe being blue shifted (cosmic ray shifted) as this information enters another rapidly approaching universe. The internal red shifts of all universes become the external cosmic ray shifts among all other Universes.

This is still a guess, however, cosmic rays do have in our Universe, a preferential direction to their origin, which may signify that Universes come in groups or possibly more specifically come in pairs.

To get an idea of how enormous the Hyperuniverse is. Think of an infinite number of "big bangs" happening every quantum of time (5 X 10^{-21} seconds.: $m_e c^2 = hf = h/T$, $T = h/m_e c^2$) over an infinite period of time. Our complete Universe becomes infinitely teeny, in comparison. If you compare the average lifetime of a universe with a quantum of time, then there are approximately 10^{39} non-violent universes per "big bang".

What this has to do with the origin of rotation and spin.:

Assuming that the background given above is true, the author wants you to picture the centers of two Universes coming together so that their centers are coming together off-center and out of the plane so that when the two parent universes combine they combine with a spin. This is all happening prior to "the" "big bang". AS the two parent universes start spinning they then contract into a spinning neutron superstar. Now I want you to picture the rotating neutron superstar exploding in a "big bang" and thereby imparting the rotation to all the matter being spewed out from the explosion.

AS you may note the rotation originated before the "big bang".

This, the author holds, is where Our Universe got its rotational energies.

Gary Mike Colasuono

1 FEBRUARY

2001

DEAR VAL.:

I AM WRITING YOU THIS LETTER TO EASE YOUR MIND. IT SURELY *ISN'T* MY INTENTION TO DESTROY ANYTHING, MUCH LESS QUANTUM MECHANICS. IT SURELY *IS* MY INTENTION TO GIVE YOU AND QUANTUM MECHANICS AN ONTOLOGICAL PICTURE OF WHAT'S REALLY GOING ON. THIS I DO IN THIS LETTER THANK YOU VAL., SINCERELY

Gary Mike Colasuono

THE STRUCTURE AND SIZE OF A NEUTRON, AND A NEW NUCLEAR CONSTANT $\hbar_N$.:

ASSUME THE NEUTRON IS A MINIATURE HYDROGEN ATOM.:

2. DEFINITIONS.:

$m_n \equiv$ mass of neutron
$m_p \equiv$ mass of proton

m_e ≡ mass of electron at rest.
m_{re} ≡ relativistic mass of moving electron
V_e ≡ velocity of moving electron
r ≡ radius of neutron
k ≡ electrical constant
$\hbar_N$ ≡ nuclear constant

Gary Mike Colasuono

THEN.:

$$\frac{m_{re}}{m_e} = \frac{m_n - m_p}{m_e} = 2.53119424111$$

WHERE.:

1. $m_n = 1.6749543 \times 10^{-27}$ kg.

2. $m_p = 1.6726485 \times 10^{-27}$ kg.

3. $m_e = 9.109534 \times 10^{-31}$ kg.

4. $m_{re} = 2.305800 \times 10^{-30}$ kg.

AND.:

$m_{re} = 2.53119424111\ m_e$

$$= m_e \, (1 - v_e^2/c^2)^{\,-.5000\ldots}$$

SOLVE FOR v_e.:

$$v_e = .918650833823c$$

$$= 275404591.516 \text{ m/s}$$

WHERE

$$c = 299792458 \text{ m/s}$$

Gary Mike Colasuono

$$\frac{m_{re} \, v_e^{\,2}}{r} = k \, \frac{e^- \, p^+}{r^2}$$

$$r = k \, \frac{e^- \, p^+}{m_{re} \, v_e^{\,2}}$$

$$r = 1.31918 \, 291452 \times 10^{-15} \text{m}$$

$$r = 1.3191829 \times 10^{-15} \text{m}$$

WHERE.:

$$k = 8.9875517874 \times 10^9 \text{N·m}^2/\text{C}^2$$

$$q = 1.6021892 \times 10^{-19}\text{C}$$

$$mvr = m_{re}\, v_e\, r = (n + \tfrac{1}{2})\, \hbar_N\, (\tfrac{2}{3})_{n=1}$$

WHEN.:
$$n = 1 \Rightarrow \hbar_N = 8.37717965312 \times 10^{-37} \text{ joule-sec}$$

$$h_N = 5.26353721121 \times 10^{-36} \text{ joule-sec}$$

$$\text{SPIN}_{\text{Nuclear}} = \frac{1}{3}\, \hbar_N$$

$$\text{For } n=1 \quad \text{and} \quad \text{SPIN}_{\text{Nuclear}} = +\tfrac{1}{2}$$

This is 93% accurate.
λ: See page 47 and $2\pi r$.

$$h_N = 2\pi\, \hbar_N$$

RELATIVISTIC QUANTUM MECHANICS.

1. $\quad hf_o = m_o c^2$

2. $\quad f_o = \dfrac{m_o c^2}{h}$

3. $\quad f = \dfrac{m_o c^2}{h} \sqrt{1 - \dfrac{v^2}{c^2}}$

4. $\quad f' = f_o - f \quad$ (BEAT FREQUENCY)

5. $\quad f' = \dfrac{m_o c^2}{h} \left(1 - \sqrt{1 - \dfrac{v^2}{c^2}} \right)$

6. $\quad f'' = \dfrac{m_o c^2}{h} \left(\dfrac{1 - \sqrt{1 - \dfrac{v^2}{c^2}}}{\sqrt{1 - \dfrac{v^2}{c^2}}} \right)$

7. $\quad f'' = \dfrac{m_o c^2}{h} \left(\dfrac{1}{\sqrt{1 - \dfrac{v^2}{c^2}}} - 1 \right) = $ TRUE BEAT

FREQUENCY

8. $\quad \lambda = \dfrac{v}{f''}, \quad \text{f'} = \dfrac{mc^2 - m_o c^2}{h} = \dfrac{K.E.}{h}$

9. $\quad \lambda = \dfrac{hv}{mc^2 - m_o c^2} = \dfrac{hv}{K.E.} \approx \dfrac{2h}{mv}$

NOTE.: $\lambda = \dfrac{h}{mv}$ represents ½ WAVELENGTH

Gary Mike Colasuono

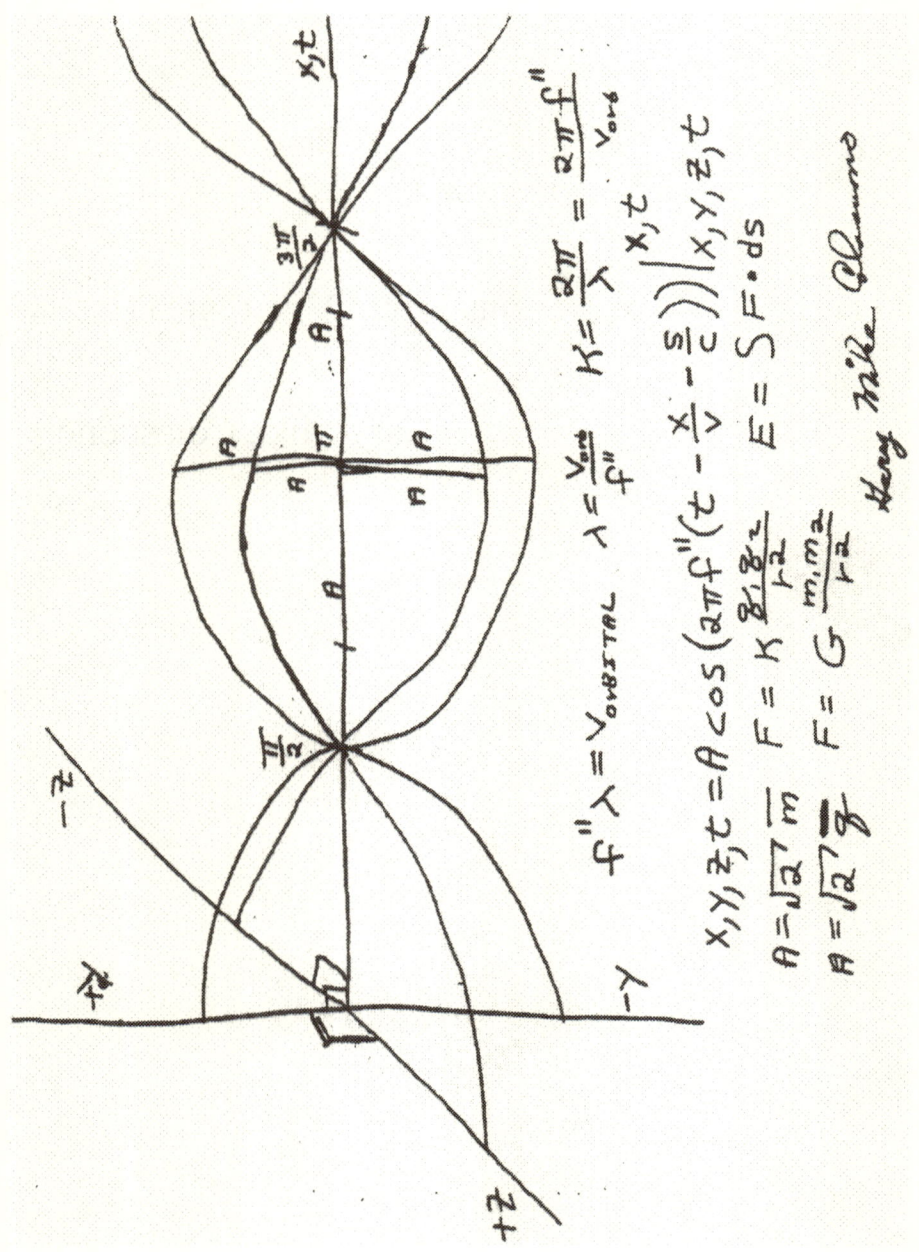

NOTES.:

1. WHAT WE MEASURE IS THE AVERAGE VALUES OF m, E, q, AND F
2. PARTICLES GO OUT OF, INTO, OUT OF, INTO, …EXISTENCE. THE PARTICLE CAN TUNNEL AT $\frac{\pi}{2}$, $\frac{3\pi}{2}$, $\frac{5\pi}{2}$,…
3. PRE 11/11/95 QUANTUM THEORY IS ONLY A METHOD OF ACCOUNTING. THIS LETTER PRESENTS THE ONLY ONTOLOGICAL PICTURE OF WHAT'S TRUE.
4. f_o AND f ARE THE HIDDEN VARIABLES. THANK YOU DAVID
5. AND THANK YOU VAL!
   SINCERELY

Gary Mike Colasuono

Gary Mike Colasuono

EMULSION BIOCHEMISTRY

ANTI-CANCER FORMULA

BY

GARY MIKE COLASUONO

Gary Mike Colasuono

Abstract.:

Emulsion Biochemistry, a new area of Biochemistry, plays a well-defined role in the prevention, halting, and curing of diseases, not the least of which is cancer.

Article.:

AS is well known among certain biochemist(s), is that both water soluble anti-oxidants (such as, but not limited to, Vitamin C) and oil soluble anti-oxidants (such as, but not limited to, Vitamins A, D, and E) are all anti-carcinogenic (anti-cancer agent(s).: that is they prevent the oxidation of compounds and elements into free radicals and they prevent the oxidation of normal cells into cancerous cells where their oxidation represents a deleterious mutation of the normal cells.).[1]

What isn't well known is Gary Mike Colasuono's suggestion that we chemically connect both the water soluble and the oil soluble anti-oxidants in an emulsion employing an emulsifier, (such as Lecithin). This then would create a chemically connected synergistic constructive drug interaction, which might make the combination much more potent than the anti-oxidants without such emulsifier. This has yet to be proven with valid experiments. Note.: There might not be a need for control groups, since past results could serve that purpose.

Furthermore, since water soluble anti-oxidants are water soluble at more than one site of the molecule, and since oil soluble anti-oxidants are oil soluble at more than one site of the molecule, therefore, the emulsion, itself, is both water soluble and oil soluble. This signifies that the emulsion would be delivered to both the water soluble and the oil soluble areas of our bodies. Thus, all types of cancer would be affected.

The formula then would be one mole oil soluble anti-oxidants to one mole emulsifier to one mole water soluble anti-oxidants.

The proceeding formula should prevent, halt, and/or cure cancer. This depends on the purity, dosages, and methods of administering the emulsion, whether orally, intramuscularly[2], intravenously[3], or intra-tumourously.

Note.:

Emulsifiers are substances whose molecules are BOTH water soluble and oil soluble, one site of the molecule being water soluble and a different site of the molecule being oil soluble.

Gary Mike Colasuono

FURTHER USES FOR EMULSION BIOCHEMISTRY.:

BY

GARY MIKE COLASUONO

Gary Mike Colasuono

Since many anti-biotics are acids or slightly water soluble salts and are thus water soluble, we could and should add one mole of emulsifier(s) to every two moles of anti-biotic(s), to create anti-biotic(s) which are both water soluble and oil soluble as are different areas of our bodies. Thus the anti-biotic(s) would then be delivered to BOTH the water soluble and the oil soluble areas of our bodies.

Note.:

The overall potential biochemical synergisms of the proceeding should be apparent.

REFERENCES:

1. Ames, Bruce N., Dietary carcinogens and Anti-carcinogens, *Science*, 1983.
2. Lerner, Nathan, private communication (intramuscularly), Hamden, CT., 2001.
3. Altman, Reinout F. A., (Rio De Janeiro, Brazil), U.S. Patent #4,252,793, Feb. 24, 1981.

PEACE OF MIND IN OUR TIME

RE.: HUMAN HUMAN BIOCHEMISTRY.:
1.: CHANGE THE EMPHASIS FROM REPRODUCTION AND BIRTH TO COUPLING AND ANNIVERSARIES AND REBIRTH BY CHANGING THE NAME OF THE REPRODUCTIVE SYSTEM TO THE COUPLING SYSTEM.: PAIR THEM YOUNG…
2.: TRADE IN ALL THE SKULL CAPS AND CROSSES FOR FIG LEAVES.: PAIR THEM YOUNG…
3.: MUCH BETTER LIVING THROUGH BIOCHEMISTRY.: PAIR THEM YOUNG.:
4.: GALATIANS CHAPTER SIX.: PAIR THEM YOUNG…

GARY MIKE COLASUONO
PROSPECTIVE
UNITED STATES SENATE

PEACE OF MIND IN OUR TIME

RE.: TAXATION AND SOCIAL SECURITY AND MEDICARE.:
1.: TRICKLE UP TAXATION IS THE ONLY VALID AND FAIR FORM OF TAXATION SINCE TRICKLE UP TAXATION EMPHASIZES NECESSITIES (FOOD, CLOTHING, SHELTER, COMMUNICATIONS, TRANSPORTATION) OVER LUXURIES, AND THEREFORE MAKES MUCH MORE SENSE.
2.: SOCIOECONOMIC PLANS:
A.: GRANT CAPITAL GAINS INCENTIVES FOR ALL THOSE INDUSTRIES WILLING TO AUTOMATE.
B.: THEN ASSIGN ALL INDUSTRIAL ROBOTS A HYPOTHETICAL HOURLY WAGE AND TAX THOSE WAGES SINCE EVERYTHING THAT CREATES WEALTH COULD AND SHOULD BE TAXED.
3.: CREATE TWO NEW CABINET POSTS.:
A.: DEPARTMENT OF BIOCHEMISTRY, AND B.: DEPARTMENT OF AUTOMATION

GARY MIKE COLASUONO
PROSPECTIVE
UNITED STATES SENATE

61

Gary Mike Colasuono

ON NUCLEAR PEACE
WITHOUT SUFFICIENT PEACEFUL TRADE,
ARMS CONTROL COULD BE INTERPRETED AS
THE RULES FOR NUCLEAR WAR

ONE WORLD
UNIFY NATO AND WARSAW PACT NATIONS
INTO ONE ORGANIZATION,
WHICH WILL THEN SERVE THE UNITED NATIONS

ON COMMUNISM
EITHER WE LEARN TO LIVE TOGETHER
OR WE WILL DIE TOGETHER

"GOD GRANT ME THY KNOWLEDGE OF PHYSICS AND BIOCHEMISTRY AND THE WISDOM TO KNOW THE DIFFERENCE."

TOWARDS A HEALTHY SOCIETY.:

By Gary M. Colasuono

In today's fast changing society, wisdom is a rare commodity. Wisdom requires free time and free thought. To acquire free time, automation can help. Today we have to make a choice, a commitment either to abandon technology and become as the strictest AMISH or develop our partially developed technology to a completely fulfilled technology synthesised with a modified AMISH philosophy. The latter can only lead to One Heaven on earth and elsewhere, A Heaven wherever we choose to go.

As in the last stanza of the Marine Corp Hymn, I feel, Heaven is the only thing worth fighting for. Adam and Eve found out that knowledge without wisdom hurt rather than helped. It was the tree of knowledge not wisdom, and God knew when they would be ready for both.

In General, creativity requires free time to develop fully. In particular creative engineers given the necessary free time can develop automation to its most complete and simplist form. This creates a social feedback loop. As automation is implemented more time and more people will become available to continue to perfect automation. The feedback loop repeats perpetually.

In the meantime, other people with newly found free time can now develop their full creative talents to perfect other aspects of the technology (space research and development), art, archectecture, interior design, entertainment skills, and so on.

The less creative can now retrain and acquire the necessary skills to acquire clean, comfortable, well paying jobs full time or part time. (note.: full time will become "part time".). The object being to always work yourself out of a job, into something much more interesting and enjoyable.

All the people in the society can now spend more time on synergetic love making (A limit of one child per new family), raising children, socialisation, meditating on their psychological well being (with or without guidance) and to perfecting the society we presently live in.

Automation also provides a hidden benefit or more appropriately a necessity for a healthy society: the separation of man and machine. Before technology, man was closer to nature and therefore more in tune with nature. A partially developed technology provides the worst condition: Man and machine mixed in factory type situations. Automation, however, provides the best alternative: factories consisting totally of machines but at the same time providing the advantages of present day technology in more abundance: Goods with standardised parts in true abundance (note.: this must be coupled with recycling (controlled thermonuclear fusion torches) to insure adequate supplies of raw materials.).

This is sort of having your angel food cake and eating good also. The above is a specific case of the more general case of the separation of hard chemistry and soft biochemistry, or stated another way the separation of electronic and mechanical devices which are hard, and biochemistry which is soft and regenerative. In most cases, the mixture of the two isn't synergetic. For instance, the mixture seems, however, done, to lead to the hellish situations: 1.: Being in a noisy factory and/or something similar: 2.: Having a factory in your head: I.e. "The Terminal Man".

In order to insure a smooth transition from the present to a fully automated and kind society we must insure a government sponsored department of retraining based on the present department of rehabilitation. The department of Retraining should require as the only qualification for eligibility the desire, on the part of the applicant, to be retrained.

It isn't only necessary on the part of the government to provide free time employing automation and thirty hour work weeks, but also provide money in the form of an Annual Guarenteed Incomes as suggested by Robert Theobald in his book entitled "Society of Abundance", to replace present day welfare, unemployment insurance, social security, veteran pensions, and scholarships. This system can also be automated and integrated into the income tax system as a "negative" income tax pro rated monthly.

The money for this system will come from what was spent on welfare, social security, unemployment insurance, veteran pensions, and scholarships as well as from the increased efficiency that automation promises. From the time the Guarenteed Incomes system is initiated (After heavy investment in Automation is achieved), the Guarenteed Incomes can be adjusted upward from some suitable threshold as the increased social efficiency of automation becomes realised. Some economic incentive should also be provided for the companies automating as well as additional economic incentive above and beyond the

Guarenteed Incomes for the layed off employees as a form of automation insurance and partial retirement plan, and decreased work hours for full time employees. For Instance, if we can provide twice as many goods with one-half the number of man hours than we should provide.: a) taxes for retaining b) thirty hour work weeks based on the present forty hour pay check. c) taxes for the Guarenteed Incomes plan if necessary d) company automation insurance e) increased profit margins f) % incentive for exceptional automation engineers, and look for a proper balance.

Mentioning efficiency in the context of social systems, the only fair and valid way of looking at efficiency is how much can be produced per man hour of work. In the limit, if machines will do all the work, a large finite quantity of goods will be produced with virtually no work on the part of humans. This leads to efficiencies approaching infinity (Given the present definition). The only other way of increasing social efficiency is to increase the level of Honesty, Directness, and Conciseness in the society.

Simultaneously, we can concentrate on increasing the physical efficiency of our machines and bring them up to somewhere in the neighbourhood of 100% efficiency (Greater than 100%, In the case of

Heat Pumps) as well as meditating on how to increase the personal happiness of each individual in the society. Abraham Maslow in his book "Toward a Psychology of Being" paves the path to realisation of the latter.

This article may sound utopian, it is; but it is also possible and practical. The only problems are the lack of practical education about automation, which this article provides, and what I call "The Law of Propagating Misery". The law of propagating misery is the present "ruling" generation's attitude towards the unhappiness in their own lives. Granted they went through misery, now its their children's "turn". Base emotion such as envy and jealousy are aimed at the advantages the younger generation will gain as a result of automation. This psychological cycle can be changed and replaced with a beautiful cycle of each ruling generation providing a superior life (youth) for each succeeding generation. This is accomplished with comprehension of the solutions and problems, wisdom, and sympathy for all mankind based on the true human condition.

This is no time for argument or debate. In fact, there is no argument. It isn't only who's correct that's important, but also who's truly happy. In most cases, however, where knowledge brings sorrow, wisdom brings True Happiness. Making the transition is worth the effort.

The purpose of this article is to make TOTAL AUTOMATION, which is practical and possible, probable.

Afterthought.: Increased production is the opposite "of" inflation and automation promises increased productivity.

NOTES.: This article presents the basis for.:

SOCIALISED BIOCHEMISTRY.:

AND.:

SOCIALISED AUTOMATION.

FURTHER NOTE.:
PAIR THE LIMIT OF ONE CHILD PER NEW FAMILY.
<u>PROTECT AND PAIR</u> THE CHILDREN INSTEAD "OF" HAVING MORE THAN ONE CHILD PER NEW FAMILY.

THE TWELVE COMMANDMENTS AND THE GOLDEN COMMANDMENT<u>S</u>.: FOURTEEN ALL.:

THOU SHALT STRIVE NOT TO COMMIT ADULTERY: THOU MAYEST SPOUSE SWAP…

THOU SHALT STRIVE NOT TO KILL.

THOU SHALT STRIVE NOT TO STEAL.

THOU SHALT STRIVE TO HONOR THY PARENTS…

THOU SHALT STRIVE TO LOVE THY NEIGHBOURS AS THYSELVES.:

"DOUBLE" DATE.: COMPOUND DATE.

THOU SHALT STRIVE TO LOVE GOD WITH ALL THY HEART, MIND, AND SOUL…

THOU SHALT STRIVE TO HAVE NO OTHER GODS BEFORE YAHWEH, MOSES AND CHRIST…

THOU SHALT STRIVE TO COLONISE SPACE.: PURPOSE…

THOU SHALT STRIVE TO CURE THE AGING PROCESS: LOVE.:

THOU SHALT STRIVE TO PAIR THY CHILDREN.: LOVE…

THOU SHALT STRIVE NOT TO CRUCIFY THY LEADER(S).: LOVE.

THOU SHALT STRIVE TO LOVE STRANGER(S) EVEN IF THEY ARE NOT OF THIS EARTH.

THOU SHALT STRIVE NOT TO DO UNTO OTHERS WHAT YOU YOURSELF DON'T WANT DONE UNTO YOURSELF. AND THOU SHALT STRIVE TO DO UNTO OTHERS WHAT YOU YOURSELF WANT DONE UNTO YOURSELF…

THE UNITED NATION'S' PLAN.:

Over one Trillion dellars U.S. per year are spent on national militaries globally. This amount could and should be redirected as follows.:

1.: 20% for a United Nation'S' Police Department. ....
2.: 20% for a United Nation'S' AeroSpace Administration. ....
3.: 20% for a United Nation'S' Department of BioChemistry. ....
4.: 20% for a United Nation'S' Department of Automation. ....
5.: 10% for middle class tax relief. ....
6.: 10% for national debts.

In addition, the United NationS could and should be organised as a weighted democracy by selling votes and vetos to the Member Nation States on a per vote per year basis and a per veto (flat rate) basis. Allso, what is understood is that every Nation State possesses a free voice and one free vote.

TRANSNATIONALL LAW.:

If and only if any Nation State is invaded and allso requests aid in repelling the invading nation, then the United Nation'S' Police Department shall repel the invading nation to pre-invasion borders *and no further*.

NOTE.:

Every Nation State possesses Nation State's' Rights as every individual possesses Human Rights as regards the following.:

1.: Federall, State, and Locall Laws.
2.: Religions.
3.: Cultures.

NOTE.:

Every Nuclear Nation State could and should turn over ten per cent per year over the next nine years of every nuclear weapon on earth to the United NationS and then turn over one per cent per year over the next ten years following the first nine years of every nuclear weapon on earth to the United NationS *until after nineteen years, the United NationS possesses a nuclear monopoly.*

Advantages of This United Nation'S' Plan.:

1.: MUCH KINDER THAN.
2.: MUCH GREATER THAN.
3.: MUCH SAFER THAN.
4.: MUCH BETTER THAN.

PURPOSE.:

TO LIVE FOR GOD AND TO LIVE FOR EACH OTHER AND TO SERVE GOD BY CREATING HEAVEN ON EARTH AS WELL AS ELSEWHERE.

GOAL.:

TO CREATE A KIND, LOVING, EFFECTIVE, STRONG, AND HONEST UNITED NATIONS AS THE VEHICLE FOR THE CREATION OF AN ALLMIGHTY FAMILY OF GODS AND GOD'S' ANGEL(S).

NOTE.:

GODS ARE FOURSOMES, THOT IS A COUPLE OF PAIRS OF GODS IN EACH CASE, AND ANGELS ARE FOURSOMES, THOT IS A COUPLE OF PAIRS OF ANGELS IN EACH CASE. (FUTURE PERFECT)…

PEACE OF MIND
IN OUR TIME: (BEYOND TIME.) (NO TIME LIMIT.).:

RE.: *EDUCATION.: (SALVATION.:* THERE'S *NO* EASY WAY, THE FOLLOWING IS THE EASIEST)

A.: CLOSE EVERY JUNIOR HIGH SCHOOL AND HAVE THE FORMER JUNIOR HIGH SCHOOL TEACHERS TEACH HIGH SCHOOL WITH THE PREVAILING HIGH SCHOOL TEACHERS, THEREBY DOUBLING THE TEACHER TO STUDENT RATIO IN THE HIGH SCHOOLS…

B.: UNIVERSITIES.: (FOURTEEN TO TWENTY.).:

   1.: SIX CONSECTUTIVE YEARS OF MATHEMATICS AS A SECOND LANGUAGE.:

      a.: FIRST YEAR.: DIFFERENTIAL AND INTEGRAL CALCULUS, VOLUME 1, BY RICHARD COURANT.

      b.: SECOND YEAR.: DIFFERENTIAL AND INTEGRAL CALCULUS,

VOLUME 2, BY RICHARD COURANT.

2.: SCIENCE (AND CONSCIENCE…).:

    a.: TWO YEARS OF PHYSICS, FIRST TWO YEARS.: FUNDAMENTALS OF PHYSICS (SECOND EDITION, EXTENDED REVISED PRINTING.: HALLIDAY, RESNICK).

THEN.: b.: TWO YEARS CHEMISTRY, SECOND TWO YEARS.

THEN.: c.: TWO YEARS BIOCHEMISTRY, THIRD TWO YEARS.

    3.: DEGREE.: MASTER OF BIOCHEMISTRY…: THE ONLY DEGREE.:

    4.: *WE LIVE IN A BIOCHEMICAL ECOSYSTEM…*

GARY MIKE COLASUONO
PROSPOCTIVE
UNITED STATES' SENATOR

NOTE.: RE.: THE *UNITED NATIONS*.: THE ABOVE IS A GLOBALL.: SUGGESTION…

REFERENCES:

1 Ames, Bruce N., Dietary carcinogens and Anti-carcinogens, *Science*, 1983.
2 Lerner, Nathan, private communication (intramuscularly), Hamden, CT., 2001.
3 Altman, Reinout F. A., (Rio De Janeiro, Brazil), U.S. Patent #4,252,793, Feb. 24, 1981.

Gary Mike Colasuono

APPENDIX A
DIMENSIONAL ANALYSIS

1) $f \equiv \dfrac{\text{EARTHSHIP CYCLES}}{\text{EARTHSHIP SECONDS}}$

2) $f^1 \equiv \dfrac{\text{SPACESHIP CYCLES}}{\text{EARTHSHIP SECONDS}}$

3) $T \equiv \text{EARTHSHIP SECONDS}$

4) $T^1 \equiv \text{SPACESHIP SECONDS}$

5) $f^1 = f\sqrt{1 - \dfrac{v^2}{c^2}}$

6) $\dfrac{f^1}{f} = \sqrt{1 - \dfrac{v^2}{c^2}}$

7) $T^1 = \left(\dfrac{f^1}{f}\right)T$

8) $T^1 = T\sqrt{1 - \dfrac{v^2}{c^2}}$

9) $f_{ss} \equiv \dfrac{\text{EARTHSHIP CYCLES}}{\text{EARTHSHIP SECONDS}}$

10) $f^1_{ss} \equiv \dfrac{\text{SPACESHIP CYCLES}}{\text{SPACESHIP SECONDS}}$

11) $\dfrac{f_{ss}}{f^1_{ss}} = \dfrac{\dfrac{A - \text{Cycles}}{1}}{\dfrac{A\sqrt{1 - \dfrac{v^2}{c^2}}\,\text{cycles}}{1\sqrt{1 - \dfrac{v^2}{c^2}}}} = 1$

12) $f_{ss} = f^1_{ss}$ *Note: ss = Same System*

APPENDIX B
ABSOLUTE MASS

1) $$\frac{m_0}{\sqrt{1-\frac{v^2}{c^2}}} = m$$

2) $$m_0 = m\sqrt{1-\frac{v^2}{c^2}}$$

3) $$\frac{m_0}{m} = \sqrt{1-\frac{v^2}{c^2}}$$

4) $$(\frac{m_0}{m})^2 = 1-\frac{v^2}{c^2}$$

5) $$\frac{v^2}{c^2} = 1-(\frac{m_0}{m})^2$$

6) $$v = c\sqrt{1-(\frac{m_0}{m})^2}$$

7) $$v^1 = c^1\sqrt{1-(\frac{m_0}{m})^2}$$

7)A) $$v^1 = c^1\sqrt{1-(\frac{m}{m^1})^2}$$

8) $$c^1 = \frac{v^1}{\sqrt{1-(\frac{m_0}{m})^2}}, m \neq m_0$$

$c^1 = c$ when $m = m_0$

APPENDIX C
ADDITION OF RELATIVISTIC VELOCITIES:

$$\mathbf{KE_1 + KE_2 + PE_0 = KE_T + PE_0}$$

1) $m_0 c^2 (1 - v_1^2/c^2)^{-\frac{1}{2}} + mc^2 (1 - v_2^2/c^2)^{-\frac{1}{2}} - mc^2 = m_0 c^2 (1 - (v_1 + v_2)^2/c^2)^{-\frac{1}{2}}$

2) $m_0 c^2 (1 - v_1^2/c^2)^{-\frac{1}{2}} (1 - v_2^2/c^2)^{-\frac{1}{2}} = m_0 c^2 (1 - (v_1 + v_2)^2/c^2)^{-\frac{1}{2}}$

3) $1 - (v_1 + v_2)^2/c^2 = (1 - v_2^2/c^2)(1 - v_1^2/c^2)$

4) $1 - (v_1 + v_2)^2/c^2 = (1 - v_1^2/c^2 - v_2^2/c^2 + v_1^2 v_2^2/c^4)$

5) $(v_1 + v_2)^2 = v_1^2 + v_2^2 - v_1^2 v_2^2/c^2$

6) $v_1 + v_2 = (v_1^2 + v_2^2 - v_1^2 v_2^2/c^2)^{\frac{1}{2}}$

NOTE: This is a conservation of mass and energy analysis

APPENDIX D
WHY A FINITE SPEED OF LIGHT AND QUANTA OF MOTION

1) Equilpartition of Space Motion and Time Motion

2) Motion is much more fundamental than velocity

3) Speed = $\dfrac{\text{Space Motion}}{\text{Time Motion}}$

Quantised $(1\delta \ll 1)$

That is:

$$\frac{0}{1} = 0 \quad \text{Velocity}$$

$$\frac{\delta}{1+\delta} \approx \delta \quad \text{Velocity}$$

$$\frac{2\delta}{1+2\delta} \approx 2\delta \quad \text{Velocity}$$

$$\frac{\infty\delta}{1+\infty\delta} \approx c \quad \text{, A Finite Number}$$

.

APPENDIX E
HOW PARTICLE PULSATION IN EXISTENCE IS POSSIBLE:

1) Positive and negative charges are 90° out of phase: that is as the negative charges go into existence, the positive charges go out of existence and vice versa: that is the process is 100% elastic: (termed: inertial reactance.)

2) Mechanism (continued): Energy exchange: the outgoing energy (that is the outgoing expanding spherical wave) focuses the incoming energy to a point of maximum existence.

3) The energy is in reality a pulsation in the curvature of space and time.

4) Since positive and negative charges are 90° out of phase they can't exist at the same point at the same time, unless they are matter antimatter in which case they are in phase, and time reverse.

NOTES:

NOTES:

NOTES:

NOTES:

NOTES:

NOTES:

NOTES:

NOTES:

NOTES:

NOTES:

NOTES:

NOTES:

ABOUT THE AUTHOR

He received an 800 on his math achievement SAT test and a 788 on his physics SAT achievement test.

He started his university education in the Unified Honors Program at the Polytechnic Institute of Brooklyn, presently the polytechnic Institute of New York. Then after his freshman year in the U.H.P., took junior level physics and math in his sophomore year at the University of Utah in Salt Lake City. He presently has five years of universities behind him in physics, chemistry, and mathematics, with many A+'s in physics and math. He allso went to the University of California, SC, and to the University of Vermont, and to Southern Connecticut State University.

www.ingramcontent.com/pod-product-compliance
Lightning Source LLC
Chambersburg PA
CBHW022025170526
45157CB00003B/1355